To Krista—
Happy Journey!

SEASONS

One Woman's Journey

Written and Illustrated
by
Dorine Rivers

River City Publishing
River City, Inc.

First Edition

Printed in Hong Kong

Rivers, Dorine. Seasons: one woman's journey/written and illustrated by Dorine Rivers.
Library of Congress Control Number 2003091821
ISBN 0-9728322-9-7

Published by
River City Publishing,
a subsidiary of River City, Inc.
Website: www.rivercityinc.com

Book design by Jeni Baird

Human Fetus photo used with permission from Petit Format/Photo Researchers, Inc.

FOR MY CHILDREN

JASON
EMILY
JOSHUA
JACOB
JARED

I'LL LOVE YOU FOREVER.

SEASONS

One Woman's Journey

SPRING

She lies in the stillness of a spring morning
curled and content and cozy.
The water is a warm blanket
wrapping her comfortably.
Sleep drifts ethereally
in and out,
in and out,
breathing effortlessly.

Then suddenly it begins—
the draining
the pressure
the tugging and pulling.

As she slides into the harsh light she cries
and she wonders,
How did I get here?

SUMMER

The end of the summer is hot
and her bangs are cut
and her eyes are bright with intrigue and excitement
and she skips alone to the crosswalk.

She smells the lavender,
the weighted peach tree
and the wild honeysuckles.
She plucks an orange trumpet,
sucks out the honey nectar
and with the flower kazoo she plays
"The Farmer in the Dell."

She sits in the sweaty classroom,
her legs fidget as she sucks on her lower lip.
She draws and paints and naps—well, sort of naps—
and eats graham crackers with her new friends
and blows foaming cumulous clouds in her milk
with a paper straw.

The teacher tells her mother she is smart,
but rambunctious,
and she colors her own birthday cake purple.

FALL/WINTER/SPRING

Elbows propped, leaning on the desk
she listens to thunder clapping wildly,
watching the cracks of lightening,
while licking the end of her pencil.

She tugs at the hem of her cheerleader skirt,
then begins her essay on Darwin's theory.
She snickers to herself about the night before,
how her mother had bought into the flat tire story.
She will never be that gullible
and her kids will never do dumb things.

The bell blasts a few short rings.
She marches in the procession,
chewing gum, blowing her biggest bubble.
She wears nothing underneath her gown
just because.
The fat principal shakes her hand,
passing her a diploma.

She walks off the stage and into the hallway
where she hands in her cap, but not her gown
and her girl friends giggle
and her guy friends wink
and the mantle of adulthood settles on her like heavy armour
and she wonders,
How did I get here?

SUMMER/FALL

The end of the summer is hot
and her bangs are cut
and her eyes are bright with intrigue and excitement.
Secret handshakes
and silly songs
ritualistic candles passed along.
Sorority houses, fraternity pranks,
she tries them all
not knowing why.

Changing her major eight times
wanting to be everything
but feeling like nothing.
Coming home,
leaving home,
and occasionally leaving herself.
Searching
Finding
Lost
Found
Knowing not what tomorrow will bring.
Can't wait to see!

WINTER

The design and the shine of the snow left behind
form patterns of winter land bliss.
She walks on air suspended
by glee
by wonder
by naivety.

Her mother cries, her father cries,
her best friend cries out of self-pity.

She kneels at the alter
and her palms are sweaty
and her eyes are teary
and her heart is racing.
Underneath the wedding gown her legs fidget
as she sucks on her lower lip beneath her veil.

Her mother says she is
a rambunctious bride,
and barefooted
she cuts the five-tiered purple cake.

SPRING

She rocks and rocks,
an old girl in a young woman's body.
She watches a rainbow crescendo into
violet
indigo
blue
green
yellow
orange
red
When her son whimpers slightly
she moves him softly to her breast.

Stroking the tiny cheeks, she studies
each little eyelash,
the perfect nose,
the lips pursing
and twitching
and involuntarily smiling.
She smells his hair,
touches each tiny finger,
then every teeny nail.

"I'll love you forever," she says.

She stares at herself in his face and wonders,
How did I get here?

SUMMER

The end of the summer is hot
and her daughter's bangs are cut
and her eyes are bright with intrigue and excitement
and they skip together to school.

She and her daughter smell the lavender,
the weighted peach tree
and the wild honeysuckles.
She shows her child how to pluck an orange trumpet,
suck out the honey nectar,
then play "The Farmer in the Dell."

Walking home alone,
she feels a hint of autumn in the morning air,
and longs for books to carry
and new clothes
and new friends
and new things to learn
and old fears to face.
She sits deserted at the kitchen table
and blows foaming cumulous clouds in her milk
with a plastic straw.

FALL

She buys a backpack so she'll fit in
but her teenage kids tell her,
"Don't even try Mom,
everybody will know you're old."
Still, she goes to class.
She doesn't remember studying this much
or getting so many A's.
Her children stick her report cards to the refrigerator
and offer her rewards if she graduates first in her class.
The bell blasts a few short rings.
She marches in the procession
and the fat dean shakes her hand,
passing her a diploma stamped with Ph. D.

Her kids whistle and cheer, and her face glows cherry red.
She wonders which is better:
Dr.
Mrs.
or
Mom
?

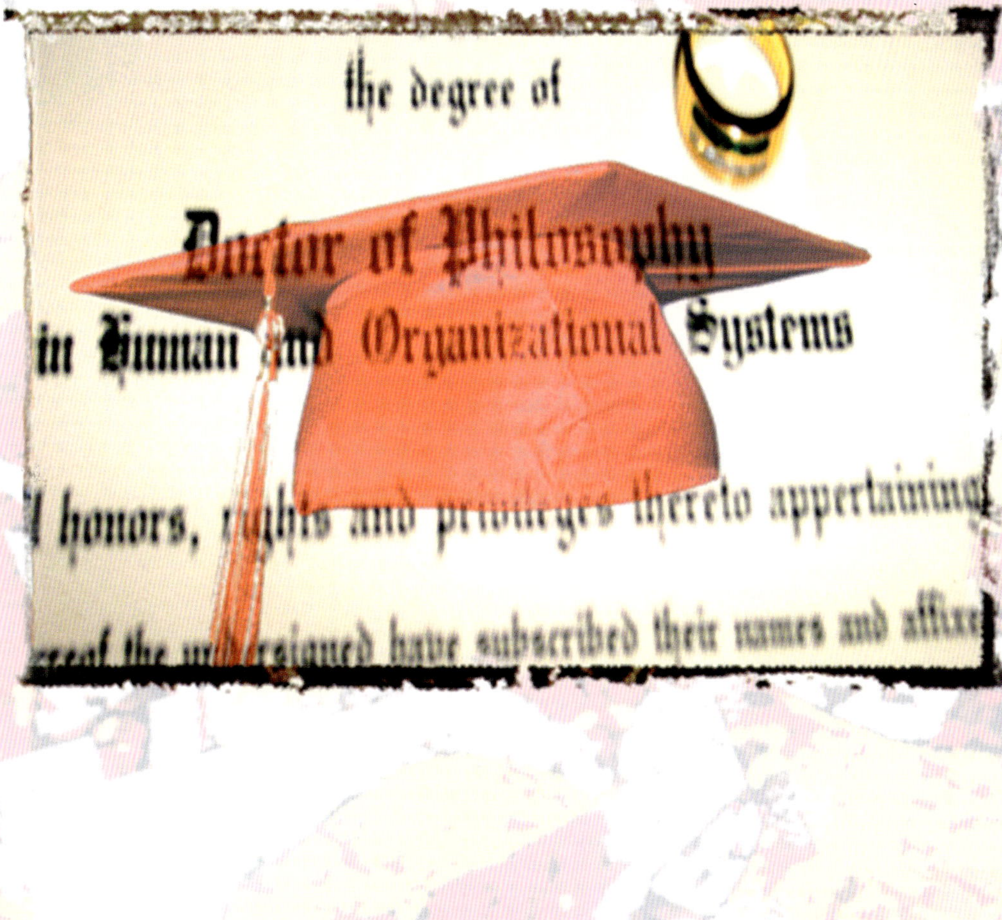
the degree of
Doctor of Philosophy
in Human and Organizational Systems

WINTER

She found them in his trunk.
She had only wanted the jack,
not a change of lifestyle.

As she reads the letters
her hands shake violently
her eyes melt the wordstogether
her brain bursts into flames
and
her heart stops breathing.

She wants him to say, "Oh, it means nothing."
Instead, he quietly packs his bags and leaves.

She busies herself unwrapping the dishes,
places them into cupboards—
which ones she doesn't care—
and shreds the moving boxes into fake confetti.

At midnight she lies backward on the frameless bed,
one mattress on the floor,
in a room she doesn't want to know.
She can hear her children crying.
Her own heaving sobs make her head hurt
and her neck
and her heart
and she cries aloud,
How did I get here?

ROAD
CLOSED

SPRING

It feels like prom
and her dress is silk
and her hair has curls
and her face is radiant
and her kids are quiet.
"Wow,"
they finally say, and
"Ooh la la."
Then the whole occasion is brought down to earth with,
"It's only a date, Mom."

He strokes her hair
and sends her a dozen purple roses
and plays with her kids
and takes them all to Six Flags
and presents her with a huge diamond ring
and says, "Do you like it?"

With the words "step" and "family" glued together
her kids ask, "Should we call him Dad?"
Her family asks, "Is he rich?"
When discussing last names her friends ask,
"Are you hy-phe-nat-ed?"

As she giggles and laughs
and listens to him sing in the shower
she smiles and wonders,
How did I get here?

SUMMER

The end of the summer is agony
and the swamp cooler's broken
and her sweat-plastered hair pokes her eyeballs
and perspiration rolls down both cheeks
and she forces herself to drive to the rehabilitation center
for gangs.

Her son had said, "Don't bother," and he hadn't cried—
at least not in front of her.
Now his eyes are puffy and blood-shot and belligerent
and he walks with a sagging amble.
"How's it going?" she asks.
Her son shrugs his shoulders stiffly.
"Okay?" she prods.
He nods painfully.
"I brought you Double Stuff Oreos," she says, smiling laboriously.
He nods awkwardly.
She lays them on the table.
He doesn't look up.
She reaches for him, stroking his cheek.
She studies each eyelash,
the perfect nose,
the lips pursing
and twitching
and involuntarily smiling.
She smells his hair,
touches each finger,
then every nail.

"I'll love you forever," she says.

FALL

One father on each arm, her daughter walks on air, suspended
by glee
by wonder
by naivety.
"I'll be happy forever!" the daughter exclaims
as she glides exquisitely by
and the mother's heart twitches
and bleeds
and mends
and she smiles.
Her daughter clutches a bouquet of white roses
and lilies
and stargazers
and baby's breath.
The silk and sequin of her white dress spin into beads and bodice,
and long-ago hours and days whirl into ancient months and years
and the words "deja" and "vu" merge together
as she gazes at her daughter and thinks to herself,
"You are me."
Now she is the mother of the bride
instead of the bride with her mother
and she wonders,
How did I get here?

WINTER

The lump is nothing.
Really.
Who needs two, anyway?
She was going to cut her hair short,
and the chemo just makes the decision easier.
After a grueling nine months her husband says, "Let's fly this coop."

So the lovebirds
become snowbirds
and leave their nest
and the state
and practically the nation.

They play the alphabet sign game, the license plate game
and count four-legged animals
and she says, "Don't count the sheep,
they might put you to sleep."
They hold hands, laugh at each other's jokes
and buy silky things at Victoria's Secret.
He's driving, she's Miss Daisy
and the maps are tattered and torn.
Seattle? Chicago? New York?
She can't tell any more.
If she knew, she could wonder,
How did I get here?

SPRING

Stroking the tiny cheek, she studies
each little eyelash,
the perfect nose,
the lips pursing
and twitching
and involuntarily smiling.
She smells her hair,
touches each tiny finger,
then every teeny nail.

Glancing up at her daughter
she compares the details
and sees the only difference is about 120 pounds.

Moving into a new role,
a new name,
a new mind set,
she is both awed and alarmed.
I am my mother, she thinks.
We are the same now.
And she wonders,
How did I get here?

SUMMER

The end of the summer is hot
and her bangs are cut
and her eyes are bright with intrigue and excitement.
She wears sunglasses that squirt ten feet
and a striped beanie with a spinning whirligig.
She rides a Harley Davidson to the soccer games
and her grandkids point and brag.

They tell her about the flat tire last night,
and she helps them think up a better story.
Passing out multiple candy bars
she invites them over for video games.

She plays poker—but not strip—until midnight,
dancing disco until three.
Late morning she's really cookin'—
chocolate waffles with whipping cream.
"Come over anytime," she tells her grandkids as they leave,
then collapses onto the sofa at noon for a nap.

FALL/WINTER

The leaves are dry
but her eyes are wet
and tired
from salt
and tears
and grief
as she walks alone to the gravesite.

She sits on the grass in the autumn sun,
her legs fidget
as she sucks on her lower lip.

She naps—well, sort of naps—
but wakes up weeping.
She sets down a dozen purple roses,
and kisses the ground where his face should be.

She reads old letters and laughs at greeting cards
whose messages and words still feel new.
The stillness makes her ears ring
while she sits quietly alone. . . listening.
She wonders,
How did I get here?

SPRING

She roams Paris and peruses Rome,
Italian cooking lessons both north and south.
She sends old postcards from
New Zealand
New Guinea
and
New Foundland
and buys china in China.
At her birthday party her kids say,
"Boy Mom, you've really put on some miles."
And then it turns into
"Got some mileage on you, baby,"
"Been around, honey,"
and
"High Flyin' Momma, won't you please come home?"

And the kids admit they all lied about the flat tire
and she says, "I know,"
and they wonder what else she knows
and she wonders what doesn't she know.

SUMMER

The end of the summer is hot
and her bangs are cut
and her eyes are bright with intrigue and excitement
and she hobbles alone to the crosswalk.

She smells the lavender,
the weighted peach tree
and the wild honeysuckles,
but she is too tired to pluck an orange trumpet
and suck out the honey nectar
and play "The Farmer in the Dell."

She yanks up her purple pantyhose,
sucks in her gut,
pinches each cheek twice
and pops in a chunk of very berry bubblegum.

She chuckles to herself that they used her suggested theme
"Old Age Ain't for Sissies"
and tries to walk without limping underneath the banner
"Welcome East High 60 Year Reunion."

She counts.
Twenty-four out of one hundred forty-nine.
"Those poor pansies that are pushing up daisies," she quietly laughs,
and smoothes back her new hairdo.
There's snow on her rooftop,
but a fire inside,
and she dances barefoot until three.

FALL/WINTER

Thunderous beads of water crash against the window
SLAM!
Roll, drip, drop, gone.
She presses her limp face against the cold glass,
feeling the drip, drip, dripping, dri-i-i-i-i-p-p-i-i-i-ing
hours, days, years, a lifetime.

Penetrating chill,
sick, better, ill.
She stares through panes and pains,
rain rolling, rain slamming, mind clamming.
No wrap around her shaking shoulders,
she threw it off weeks ago.

She stares at the wintry beads
now forming into flakes, frail and flimsy.
She rubs her bony fingers, searching for warmth
and comfort
and youth.
Haunted by both past and future
she wonders,
How did I get here?

SPRING

She lies in the stillness of a late spring night,
a young woman in an old lady's body.
Her blanket is warm like bath water
wrapping her comfortably.
Sleep drifts
in and out,
in and out,
breathing erratically.

Then suddenly it begins—
the draining of consciousness
the pressure of her heart
the tugging and pulling,
a whirling ride in a spinning tunnel.
As she slides into the light she is amazed
at the warmth and peace and beauty.
Sun fills the room.
A vibrant rainbow stretches amazing colors across the sky.
She squints her eyes, blending the seven colors:
Imperial Purple
Prussian
Emerald
Turquoise
Chartreuse
Vermillion
Fuchsia
A gentle, strong hand reaches for hers
and she no longer wonders,
How did I get here?

The End